Adding and Subtracting at the Lake

By Amy Rauen

Illustrations by Lorin Walter

Reading consultant: Susan Nations, M.Ed., author/literacy coach/consultant in literacy development

Math consultant: Rhea Stewart, M.A., mathematics content specialist

WEEKLY READER®
PUBLISHING

Please visit our web site at www.garethstevens.com
For a free color catalog describing our list of high-quality books, call 1-800-542-2595 (USA) or 1-800-387-3178 (Canada). Our fax: 1-877-542-2596

Library of Congress Cataloging-in-Publication Data

Rauen, Amy.
Adding and subtracting at the lake / Amy Rauen.
p. cm. — (Getting started with math)
ISBN-13: 978-0-8368-8983-3 (lib. bdg.)
ISBN-10: 0-8368-8983-5 (lib. bdg.)
ISBN-13: 978-0-8368-8988-8 (softcover)
ISBN-10: 0-8368-8988-6 (softcover)
1. Addition—Juvenile literature. 2. Subtraction—Juvenile literature.
I. Title.
QA115.R383 2008
513.2'11—dc22 2007026331

This edition first published in 2008 by
Weekly Reader® Books
An Imprint of Gareth Stevens Publishing
1 Reader's Digest Road
Pleasantville, NY 10570-7000 USA

Senior Editor: Brian Fitzgerald
Creative Director: Lisa Donovan
Graphic Designer: Alexandria Davis

Printed in the United States of America

CPSIA Compliance Information: Batch #CR012040GS: For further information contact Gareth Stevens, New York, New York at 1-800-542-2595

Note to Educators and Parents

Reading is such an exciting adventure for young children! They are beginning to match the spoken word to print and learn directionality and print conventions, among other skills. Books that are appropriate for emergent readers incorporate these conventions while also informing and entertaining them.

The books in the *Getting Started With Math* series are designed to support young readers in the earliest stages of literacy. Readers will love looking at the full-color photographs and illustrations as they develop skills in early math concepts. This integration allows young children to maximize their learning as they see how thoughts and ideas connect across content areas.

In addition to serving as wonderful picture books in schools, libraries, and homes, the *Getting Started With Math* books are specifically intended to be read within guided small reading groups. The small group setting enables the teacher or other adult to provide scaffolding that will boost the reader's effort. Children and adults alike will find these books supportive, engaging, and fun!

Susan Nations, M.Ed.
author/literacy coach/consultant in literacy development

I am at the lake.
It is sunny.

I see animals.

They are in and near the lake.

2

I see 2 ducks.
They swim in the water.

$2 + 2 = 4$

I see 2 more ducks.
There are 4 ducks now.

5

I see 5 bugs on a rock.

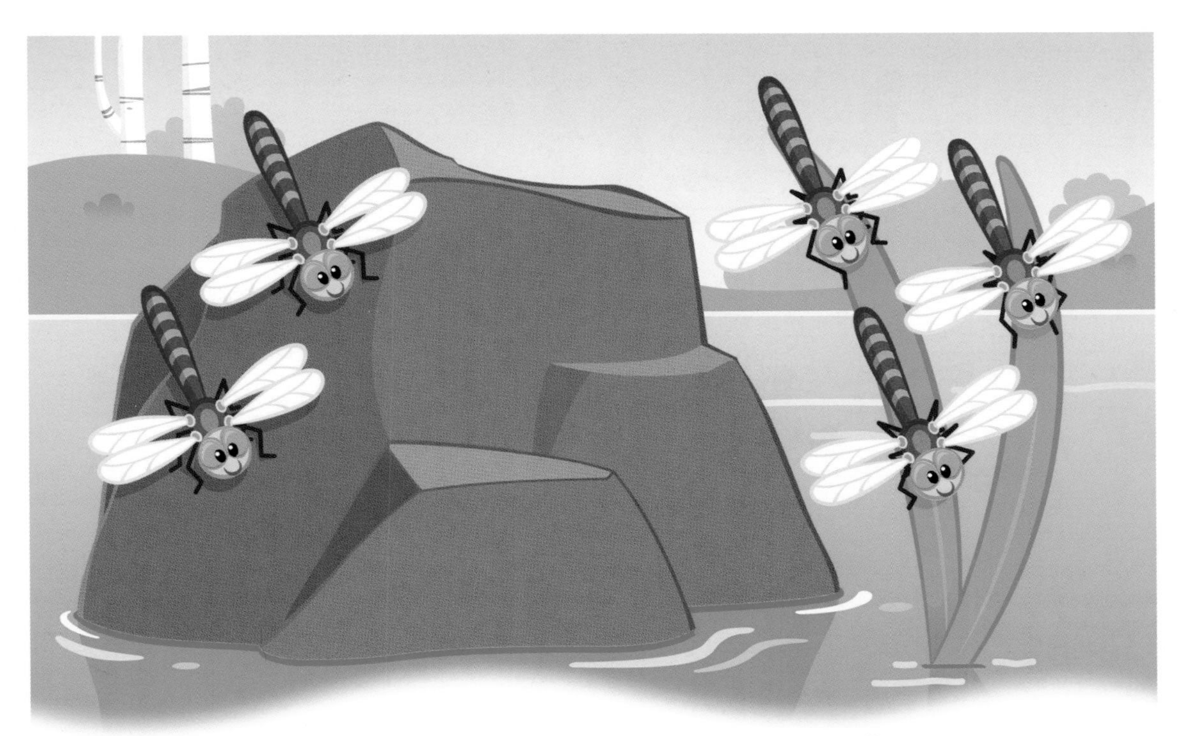

5 – 3 = 2

I see 3 bugs go.

Now 2 bugs are left on the rock.

7

I see 7 birds in the sky.

7 + 2 = 9

I see 2 more birds.
Now 9 birds are in the sky.

6

I see 6 frogs on a log.

6 − 1 = 5

I see 1 frog go. Now 5 frogs sit on the log. I like the lake!

Glossary

bird

bug

duck

frog

lake

Show What You Know

1.  swim in the lake. 
   more come.
   How many frogs are there in all?

2.  sit in a tree. 
   fly away.
   How many birds are left?

3. I see  . I see  come.
   How many bugs do I see in all?

To Find Out More

Adding and Counting On. Basic Math (series).
Diyan Leake (Heinemann Read and Learn)

Subtracting. I Can Do Math (series).
Rozanne Lanczak Williams (Gareth Stevens)

About the Author

Amy Rauen is the author of more than a dozen math books for children. She also designs and writes educational software. Amy lives in San Diego, California, with her husband and theiar two cats.